Early SCIENCE Encyclopedias

ROCKS AND MINERALS

by Meg Marquardt

Early Encyclopedias

An Imprint of Abdo Reference

abdobooks.com

abdobooks.com

Published by Abdo Reference, a division of ABDO, PO Box 398166, Minneapolis, Minnesota 55439.

Printed in China.
052024
092024

THIS BOOK CONTAINS RECYCLED MATERIALS

Editor: Arnold Ringstad
Series Designers: Candice Keimig, Joshua Olson
Production Designers: Ryan Gale, Joshua Olson

Library of Congress Control Number: 2023949601

Publisher's Cataloging-in-Publication Data

Names: Marquardt, Meg, author.
Title: Rocks and minerals / by Meg Marquardt
Description: Minneapolis, Minnesota: Abdo Reference, 2025 | Series: Early science encyclopedias | Includes online resources and index.
Identifiers: ISBN 9781098294274 (lib. bdg.) | ISBN 9798384913542 (ebook)
Subjects: LCSH: Rocks--Juvenile literature. | Minerals--Juvenile literature. | Geology--Juvenile literature. | Geoscience (Geology)--Juvenile literature. | Science--Juvenile literature. | Earth sciences--Juvenile literature. | Encyclopedias and dictionaries--Juvenile literature.
Classification: DDC 550--dc23

CONTENTS

INTRODUCTION

The world is full of different kinds of rocks and minerals.

When people walk outside, rocks and minerals crunch beneath their feet. Those rocks and minerals make up Earth's surface. Some are millions or billions of years old.

Geologists are scientists who study rocks and minerals. They have identified more than 4,000 types of rocks and minerals. Many rocks look similar. But the world of rocks and minerals is full of surprises. Some rocks are pieces of cooled lava that was shot out of a volcano. Some minerals are precious gems, such as diamonds.

Geologists collect rock samples to learn more about a region's past.

Igneous Rocks

What Are Igneous Rocks?

Earth's surface is called the crust. Below the crust is the mantle. The mantle is made of very hot rock. Its highest temperature is about 6,692 degrees Fahrenheit (3,700°C).

The mantle lies beneath Earth's solid crust.

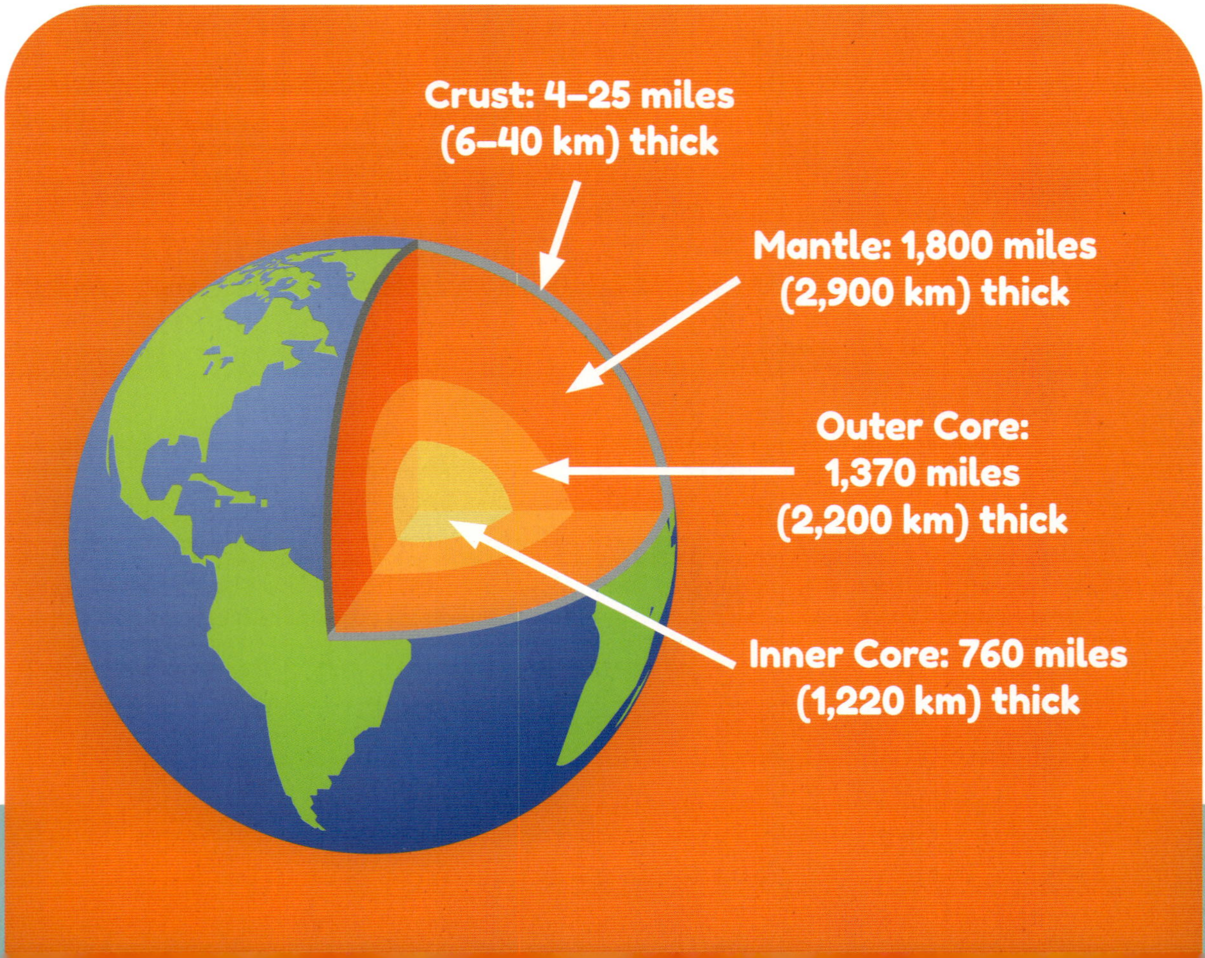

Andesite is one example of an igneous rock.

Rising from Below

Sometimes materials from the mantle make it to the surface. These materials are called igneous rocks. Igneous rocks usually take a long time to form. They are already cooled when they reach the surface. However, igneous rocks sometimes form when volcanoes erupt. Material from the mantle quickly reaches the surface.

Molten rock is so hot that it glows.

How Are Igneous Rocks Formed?

Earth has two types of igneous rocks. One is called extrusive. The other is called intrusive. Both types come from magma. Magma is molten rock. Things that are molten are liquid and hot. Magma is found under Earth's crust.

Magma and Lava

Magma slowly rises to Earth's surface. It sometimes spews from volcanoes. When that happens, magma becomes lava. Extrusive rocks are made of lava. Lava cools quickly on the surface. The fast cooling makes some extrusive rocks flat and shiny.

Cooling Matters

The same magma can form different rocks. If magma cools slowly, it may become granite. If it cools quickly, it may become rhyolite.

Magma is known as lava once it reaches the surface.

Intrusive igneous rocks are also known as plutonic rocks.

Intrusive Igneous Rocks

Sometimes magma doesn't come out of a volcano. It gets trapped below the crust. It cools slowly. This forms intrusive igneous rocks.

Forming Structures

Intrusive rocks can form large structures. That's because they form over a long time. Some intrusive rock crystals are as big as a human hand.

Igneous Rock Formation

Intrusive igneous rock cools slowly, while extrusive igneous rock cools quickly.

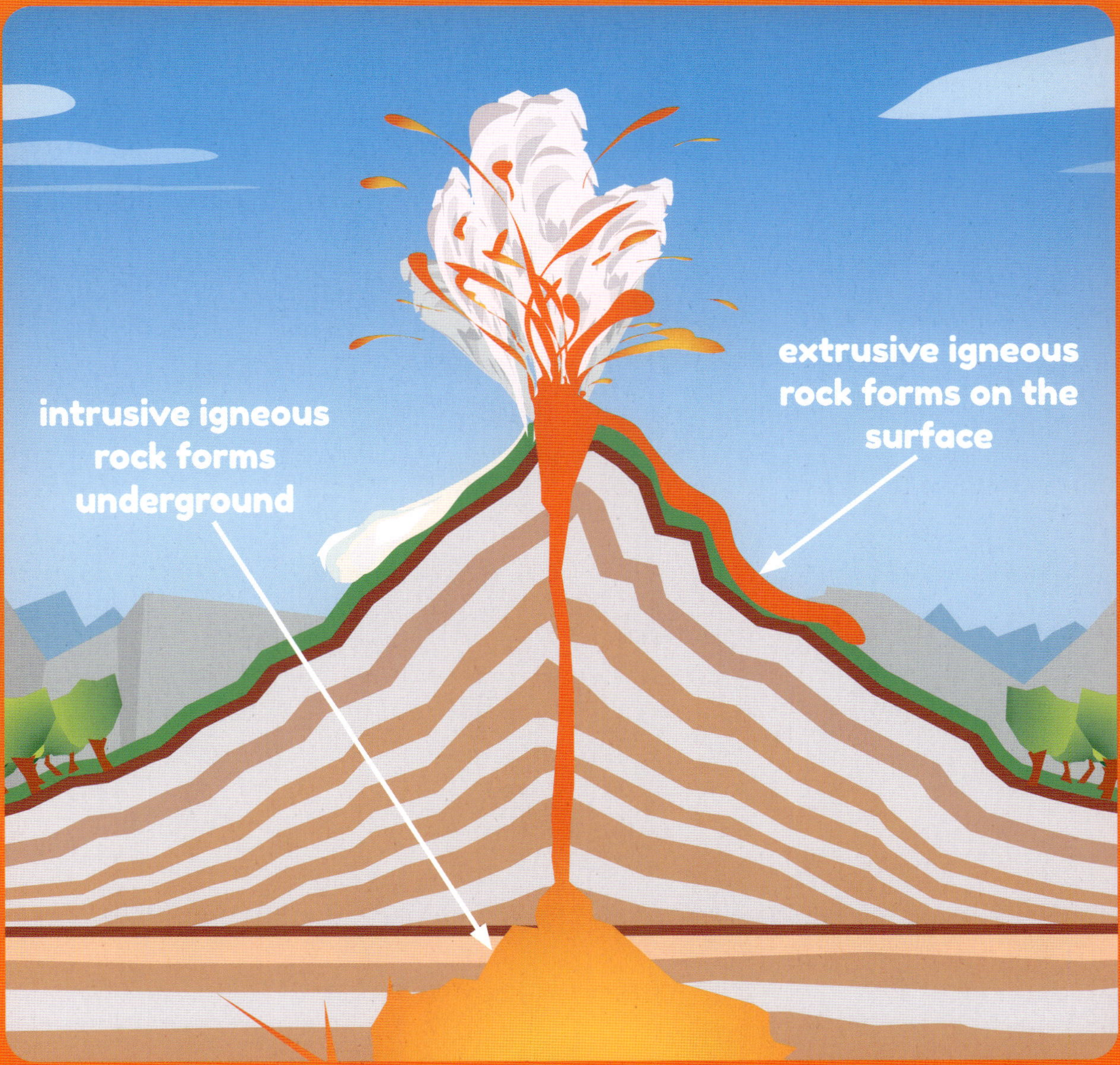

What Do Igneous Rocks Look Like?

Igneous rocks can be many shapes and sizes. If rocks are formed above ground, they might look very bumpy. They may even have small holes. These are called pyroclastic rocks. One example is pumice.

Pumice has many small holes. This allows the rock to hold so much air that it can float on water.

When obsidian breaks, it leaves very sharp edges.

Obsidian

Obsidian also forms above ground. Obsidian is made when lava cools very fast. It is black and shiny.

FUN FACT!

Some ancient arrowheads of Native American groups in the western United States are made from igneous rocks.

Geologists named gabbro after an Italian village of the same name.

Gabbro

Some igneous rocks form belowground. One example is gabbro. These rocks are dark brown or black. They sometimes contain brighter minerals such as olivine or pyroxene.

Granite

Granite also forms belowground. It is one of the most common igneous rocks. Granite is white with spots of black crystals. These crystals can be seen with the naked eye. Granite rocks can be huge. Some large monuments, such as Mount Rushmore, are carved into granite.

Workers used dynamite and drilling to create the Mount Rushmore portraits on a granite mountainside.

Where Are Igneous Rocks Found?

Igneous rocks can be found all over the world. Certain types are found near volcanoes. Tall cliffs and mountains can also be made of igneous rock.

Vermont is home to a large granite quarry.

Cooling lava becomes igneous rock.

FUN FACT!

The sea floor is mostly igneous rock from ancient volcanoes.

Lava Rocks

Many igneous rocks are made of cooled lava. They are found where lava once flowed. Geologists use these rocks to figure out the sizes of past eruptions.

Cooling Magma

After volcanoes stop erupting, magma remains below them. These large pools of magma cool under the crust. They form intrusive igneous rocks.

Red granite is one of many kinds of igneous rocks that cool underground.

Scientists think Devils Tower in Wyoming was exposed 5 to 10 million years ago.

Exposed Igneous Rocks

When volcanoes erode, igneous rocks are exposed. They can be as big as mountains. Some of the largest are in Wyoming, California, and Australia.

Devils Tower

Devils Tower in Wyoming is a large intrusive rock. It is 867 feet (264 m) tall.

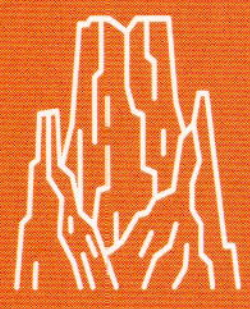

The layers in sedimentary rock formations are often clearly visible.

Sedimentary Rocks

What Are Sedimentary Rocks?

Sedimentary rocks are found on Earth's surface. They are often made of many smaller rocks. They may form in layers over time.

Layered Rocks

Sedimentary rocks can be very big. Mountains are covered with layers of these rocks. The layers can be different colors. In deserts, sedimentary rocks have red and brown layers. Sedimentary rocks can also be very small. The smallest types include sand, clay, and silt.

Layers can sometimes be seen within individual sedimentary rocks.

Sandstone forms when many layers of sand are deposited over time.

How Are Sedimentary Rocks Formed?

Sedimentary rocks form aboveground. Other kinds of rocks are created by heat. But sedimentary rocks form by deposition.

Shifting Sediment

Deposition happens when sediment is moved from one place to another. Sediment is made up of small pieces of sand, mud, or pebbles. Wind, water, and ice move it around. When sediment lands in a new place, that is called deposition.

Rivers can carry sediment to a new location.

Shale is a sedimentary rock formed from layers of mud.

Pressing Together

Sand and pebbles start out as loose pieces. As sediment builds up, all those loose pieces push down on the ones beneath. The pieces press together. Over time, this creates a sedimentary rock.

Fossils

Fossils are the remains of living things. They can be covered by sand and mud. They are often found in sedimentary rocks.

Forming New Layers

Different types of sediment can settle in the same place. In one place, most sediment might be deposited by wind.

Then a nearby lake might flood. That would bring lots of mud and pebbles. A new layer of sediment would form.

Sedimentary Rock Formation

Sedimentary rock forms when layers of sediment get more tightly packed together over time.

What Do Sedimentary Rocks Look Like?

Some sedimentary rocks are clastic. This means they are made of many pieces of rocks and minerals. Each piece is called a clast.

Natural processes often expose sedimentary rocks.

Clastic Rocks

Clastic rocks often look like many smaller rocks glued together by hard dirt. They may have different-colored layers. Their colors are usually browns and reds. The Grand Canyon in Arizona has clastic sedimentary rocks.

Sylvite is one mineral left behind by evaporation in sedimentary rocks.

Layered Minerals

Sedimentary rocks can also be made of layers of minerals. Those minerals are left behind by evaporated water. The layers can be very thin. They are often gray, white, and black.

Biological Sedimentary Rocks

Biological sedimentary rocks are made of dead organisms. Coal is one example. This black rock is made of dead plants. Limestone is made of dead sea life. It is white and porous.

Coal can be burned to release energy.

Some sedimentary rocks form on coastlines.

Where Are Sedimentary Rocks Found?

Sedimentary rocks are found on Earth's surface. These rocks are found in many different environments. A place's weather and minerals determine what sort of rock forms.

Limestone and Calcite

Limestone is a sedimentary rock formed in water. It is made of the mineral calcite. Calcite is found in the shells of animals such as clams and oysters.

FUN FACT!

Much of the United States was once covered by water.

Forming Limestone

When these animals die, their shells fall to the ocean floor. The shells form limestone. If limestone is found on land, that land was probably once underwater.

Shells and other evidence of past sea life can sometimes be seen in limestone.

Sandstone

Rivers create a different type of sedimentary rock. Rivers carry small grains of sand. The grains are pressed together over time. This creates sandstone.

Antelope Canyon in Arizona is made of sandstone.

Mudstone

Sedimentary rocks also include mudstones. Shale is a type of mudstone. It is made of clay and silt. Mudstones are typically found where water is or was.

A metamorphic rock called schist can contain various minerals.

Metamorphic Rocks

What Are Metamorphic Rocks?

Metamorphic rocks always begin as a different kind of rock. They could start as an igneous rock. Or they may start as a sedimentary rock.

Heat and Pressure

Heat and pressure turn rocks into metamorphic rocks. Metamorphic rocks are formed in Earth's crust. They can take millions of years to form. They eventually reach the surface. Gneiss, slate, and marble are all metamorphic rocks.

Heat and pressure transform shale into slate.

Limestone becomes marble through metamorphic processes.

How Are Metamorphic Rocks Formed?

Metamorphic rocks need high heat or lots of pressure to form. The process may also involve mineral-rich fluids. It often requires more than one of these things.

Smaller and Denser

Heat and pressure make igneous or sedimentary rocks smaller and denser. Fluids introduce new minerals to the rock. The result is a new metamorphic rock.

Metamorphic Rock Formation

Pressure from above and heat from below combine to make metamorphic rocks.

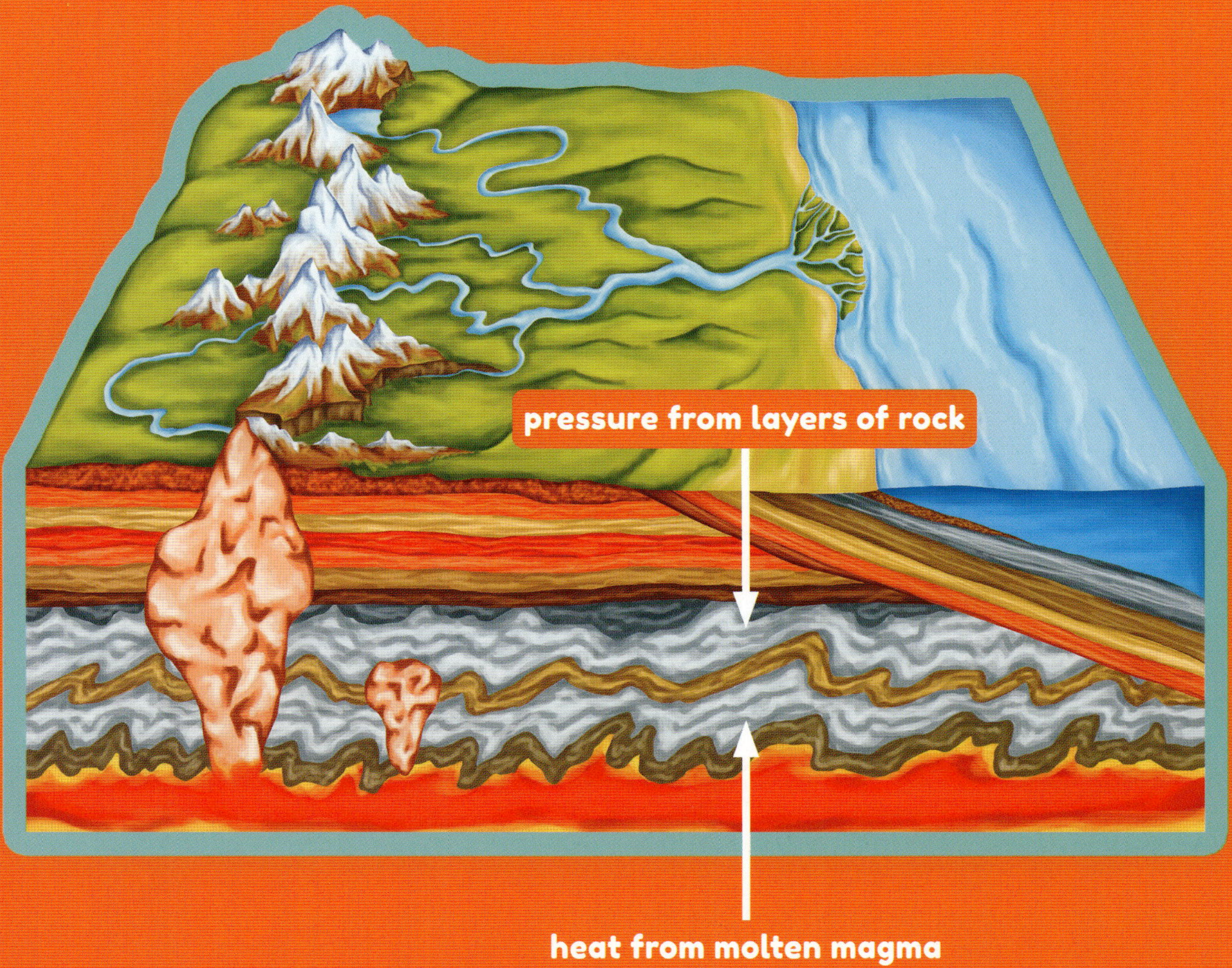

From Granite to Gneiss

Heat and pressure cause rocks to change structure. For example, granite is an igneous rock. It has spots. Under high heat and pressure, it becomes gneiss. Gneiss is pronounced *nice*. Gneiss is a metamorphic rock.

Gneiss has black lines running through it.

Rocks that melt become magma and later cool into igneous rocks.

Melting

If a rock gets too hot, it will melt. When this happens, it is no longer a metamorphic rock. A melted rock becomes an igneous rock instead.

Foliated rocks appear to have neatly stacked layers.

What Do Metamorphic Rocks Look Like?

Metamorphic rocks sometimes have an unusual look. It is called foliation. Foliation happens when minerals line up.

Stacking Minerals

Other rocks have minerals scattered throughout them. In foliated metamorphic rocks, minerals are more organized. They are stacked up like book pages. Phyllite has wavy foliation lines. Instead of breaking into flat pieces, it breaks with wavy edges.

FUN FACT!

Marble is a metamorphic rock used in art and buildings.

Phyllite may begin as shale, turn into slate, and eventually become phyllite.

Nonfoliated Rocks

Metamorphic rocks don't always have foliation. Some have solid colors. These are called nonfoliated metamorphic rocks.

Quartzite is a type of nonfoliated metamorphic rock.

Serpentine is the state rock of California.

Metamorphic Colors

Metamorphic rocks come in many colors. Serpentine is green. Marble can be white, gray, or pink. Quartzite can be purple.

Uplift moves some parts of the crust up and other parts of the crust down.

Where Are Metamorphic Rocks Found?

Metamorphic rocks start out in Earth's crust. One way they reach the surface is through uplift. Uplift happens when large pieces of the crust crash together.

Making Mountains

During uplift, parts of the crust are pushed upward. This process forms hills and mountains. Metamorphic rocks are inside those hills and mountains. Weathering and erosion expose them.

Exposed gneiss can be seen on the coast of northwest Scotland.

The Appalachian Mountains

The Appalachian Mountains are a great place to see metamorphic rocks. These mountains are in the eastern United States. They formed almost 500 million years ago. They have experienced a lot of erosion. Today, people can see many metamorphic rocks there. One type is gneiss.

The Appalachian Mountains have a rich geological history.

Marble from quarries in Italy is sold around the world.

Marble

Marble is also found near mountains. One of the world's largest marble quarries is in Vermont. It is 1.5 miles (2.4 km) deep.

The processes of weathering and erosion can create amazing rock formations.

Weathering and Erosion

What Is Weathering?

Rocks break down over time. Some rocks experience weathering. Over a long time, this process can make a rock completely disappear.

Causes of Weathering

Weathering has three main sources. First, humans and animals may cause it. Second, it can have chemical causes. Third, physical causes such as water, wind, and temperature can weather rocks.

Effects of Weathering

Weathering can cause a rock to become completely smooth. It can also create large holes in rocks. Weathering sometimes makes rocks become weak and break apart.

When holes are weathered in sandstone, it is sometimes called honeycomb weathering.

Water

Water is a powerful force of nature. It can wear down even the toughest rocks. The Grand Canyon was carved by water over millions of years.

Geologists think the Colorado River began carving the Grand Canyon about 6 million years ago.

Water repeatedly freezing and melting can crack rocks open.

Freezing

Water can fill any crack in a rock. If temperatures drop, it freezes. That pushes the crack open. The next time water gets in and freezes, the crack gets even bigger. Eventually, part of the rock might fall off.

Absorbing Water

Some rocks absorb water. That causes them to get bigger. This process weathers the rocks around them.

Large temperature swings in deserts can cause weathering in the rocks there.

Temperature

Rocks react to different temperatures. Hot temperatures make rocks expand, or get bigger. When it is cool, rocks contract, or get smaller. These changes happen over and over again. This can cause cracks in rocks. Over time, the rocks start to crumble.

Wind

Wind can also weather rocks. Wind blows around small grains of sand. That sand hits rocks. The tiny grains slowly chip away at the rocks. Over a very long time, wind can cause big changes.

Blowing sand can reshape deserts and weather rocks over time.

Pollution from power plants can lead to weathering processes.

Weathering by Humans

Humans cause weathering too. Air pollution is a major cause of weathering. When humans burn coal or gas for energy, it releases chemicals into the air. Those chemicals mix with rain to make acid rain. Acid rain eats away at rocks.

Weathering by Animals

Animals can also cause weathering. Prairie dogs are small mammals. But they can have a big impact. They dig underground living spaces called warrens. As prairie dogs dig, they slowly chip away at rocks.

FUN FACT!

A formation in Nebraska called Chimney Rock has lost 30 feet (9 m) in 150 years due to weathering.

Prairie dogs and other digging animals can weather rocks.

What Is Erosion?

Weathering breaks down rocks. Erosion happens when pieces of rocks are moved to another place. As erosion happens, rocks may change their shape and size.

Erosion can affect coastal areas repeatedly hit by waves.

Glaciers have caused widespread erosion over many years.

Causes of Erosion

Water, ice, wind, and even other rocks cause erosion. Ocean waves can reshape shorelines. Frozen glaciers can carve valleys. Blowing sand can naturally polish rocks. Rockslides can change the appearance of a mountain. These are all examples of erosion.

The Rock Cycle

What Is the Rock Cycle?

Rocks don't stay the same forever. Over time, they change into other rocks. This process is called the rock cycle.

Melting Rock

For example, a metamorphic rock might be deep in Earth's crust. Forces in the crust push the rock toward magma. This makes the rock melt.

Rocks melt and cool as part of the rock cycle.

Reaching the Surface

Melted rock is shot out of a volcano. It cools and becomes an igneous rock. This rock sits on the surface. Natural forces weather and erode it.

The Weathering Process

That igneous rock is weathered by heavy rains. The water breaks off small pieces of the rock. The water then carries those pieces away, causing erosion.

Eruptions are one violent way that rocks form.

Natural forces wear away volcanic rock formations.

Some sedimentary rocks have colorful layers.

Forming Sediment

The water drops small pieces of rock at different locations. Some of the pieces join other rock chunks. Eventually, these pieces are pressed together. They form a sedimentary rock.

FUN FACT!

Mount Everest, the world's highest mountain, is made mostly of metamorphic and sedimentary rock.

Becoming Metamorphic

This sedimentary rock is buried under more and more sediment. Pressure underground turns it back into a metamorphic rock. Then the rock cycle starts again.

The Rock Cycle

The rock cycle explains the relationships between different types of rocks.

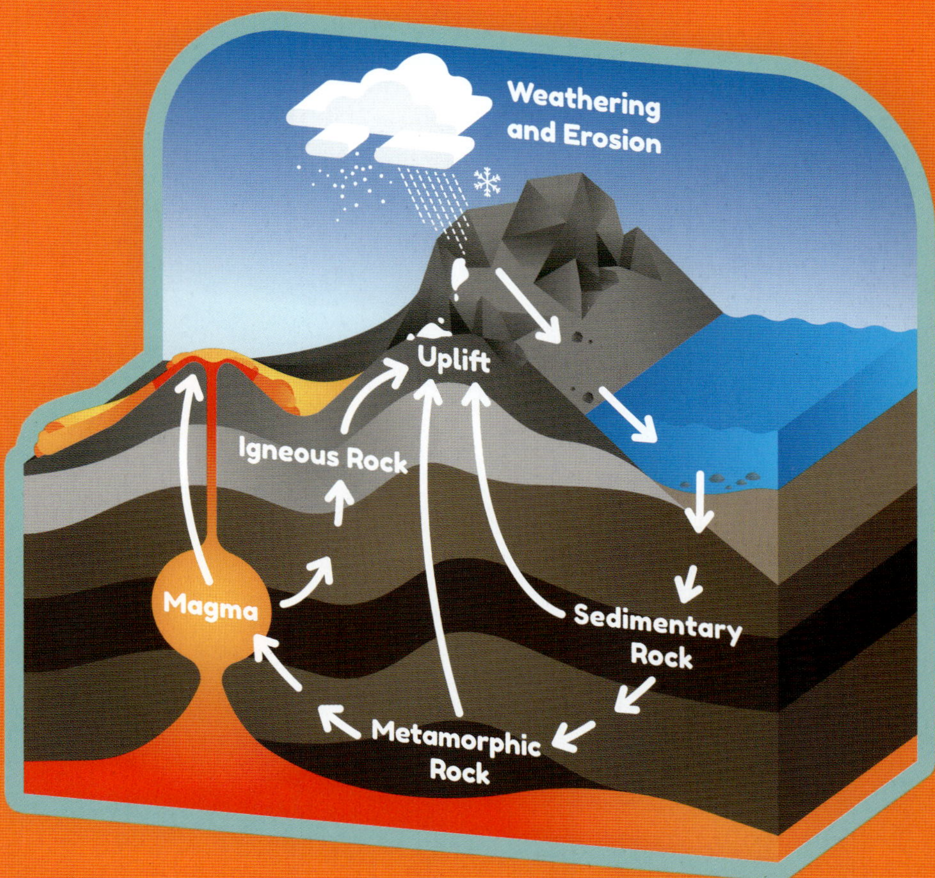

Tectonic Plates

Earth's crust is not one big piece of rock. It is broken into many pieces. These pieces are called tectonic plates. Tectonic plates drive the rock cycle.

The North American tectonic plate contains most of the North American continent.

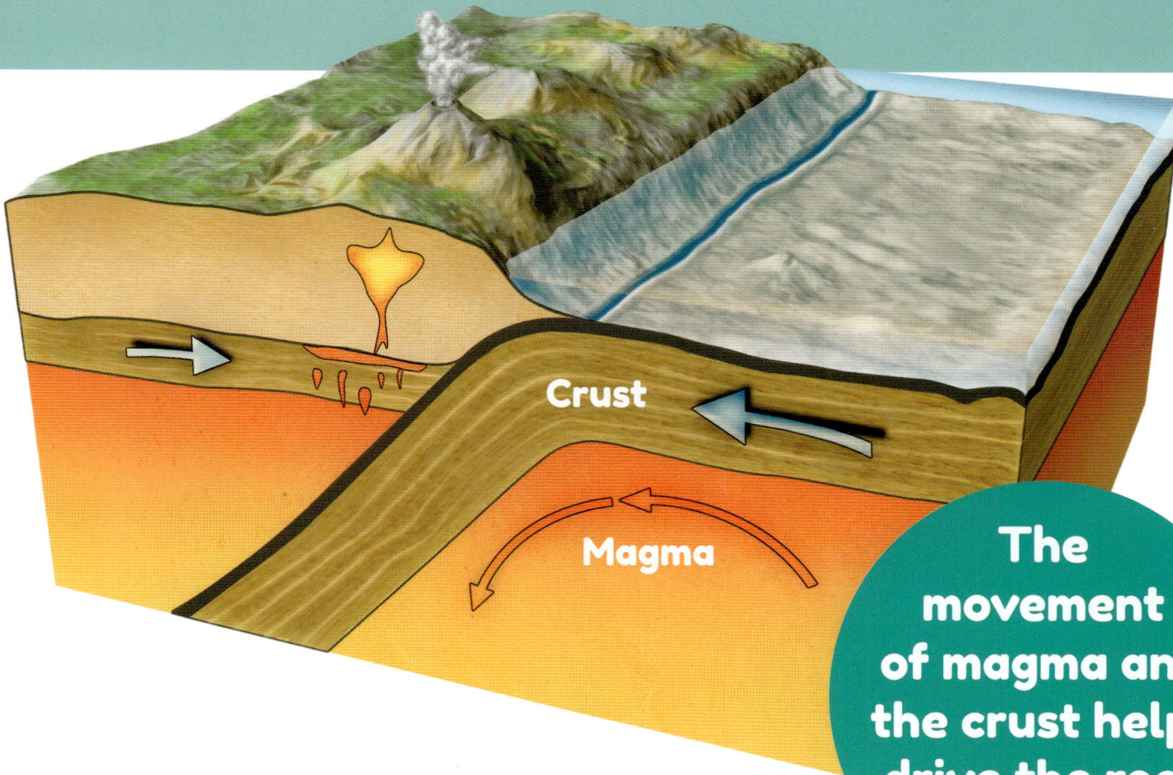

The movement of magma and the crust helps drive the rock cycle.

Uplift

The plates drift on top of the mantle. They move and collide. This can cause earthquakes and uplift. When the plates collide, they sometimes push the crust upward. This creates hills and mountains. The process moves rocks through the rock cycle.

Pangea

The continents move and shift over time. They were once joined together to form one huge continent. It was called Pangea. Pangea existed about 300 million years ago.

The water cycle is closely linked to both the rock cycle and life on Earth.

Water and the Rock Cycle

Water also has a cycle. Water on the ground evaporates into the air. That water falls again as rain. Then it evaporates again. This cycle is linked to the rock cycle.

Gathering Minerals

Rainwater can contain minerals such as salt. The water collects those minerals when traveling through rocks. By removing these minerals, the water can break down rocks.

Evaporation and New Rocks

Minerals also create rocks. When water evaporates, it sometimes leaves minerals behind. These minerals build up over time. They create rocks such as limestone and flint.

Evaporation leaves behind minerals, such as salt.

What Are Minerals?

Minerals are found in rocks. A rock might contain more than one mineral. Different minerals form in different parts of the world.

There are many kinds of minerals.

Some minerals have a colorful, shiny appearance.

Characteristics of Minerals

Minerals have five main characteristics. First, they are found in nature. Second, they are solid at room temperature. Third, they are inorganic. That means they are not made from living things. Fourth, minerals usually form crystal structures. Fifth, each type of mineral is always made of the same chemicals.

How Are Minerals Formed?

Minerals always form in an ordered structure. Like all things, minerals are made of atoms. In some materials, atoms form random patterns. In minerals, the atoms are ordered. Crystal structures are the most common. Crystals have repeating patterns.

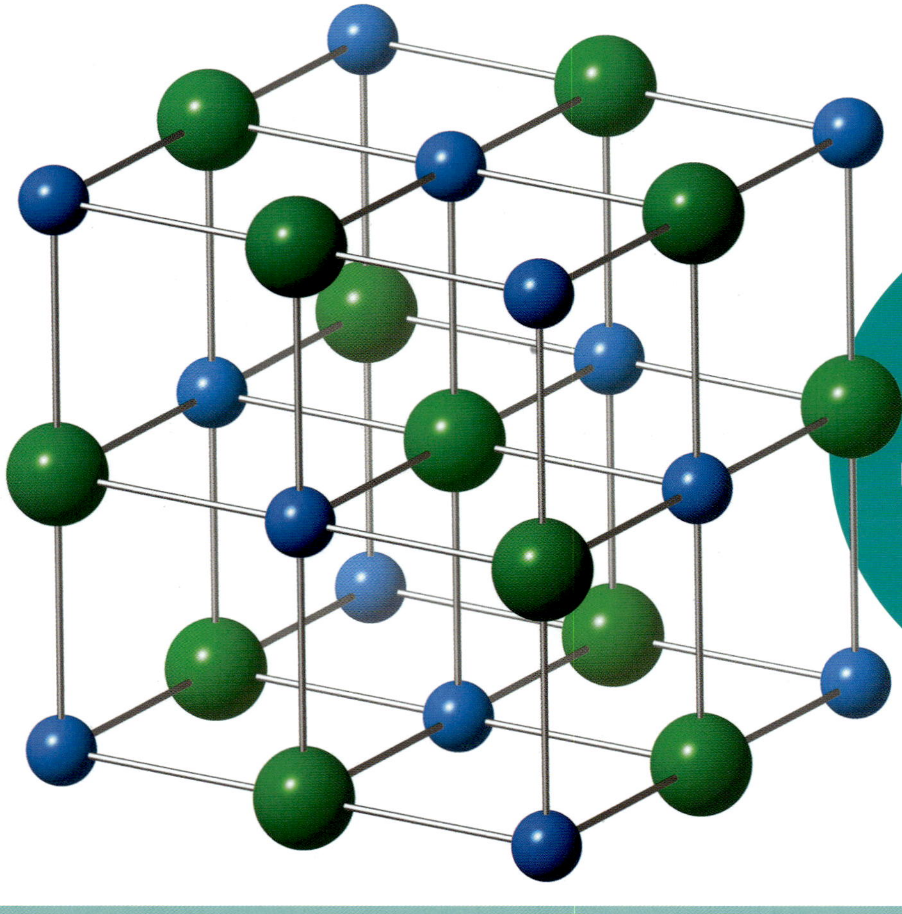

The arrangement of atoms in a mineral affects the mineral's properties.

The mineral galena breaks off in cube shapes because of its crystal pattern.

Crystal Patterns

Crystal patterns differ among minerals. Two minerals might contain the same atoms. But their atoms are in different patterns. This means they are different minerals. For example, graphite and diamond are both made of carbon atoms.

Cleavage

Some minerals have weaknesses in their structure. They break easily along those weak points. The way these minerals break is called cleavage. They may break in sheets, cubes, or other shapes.

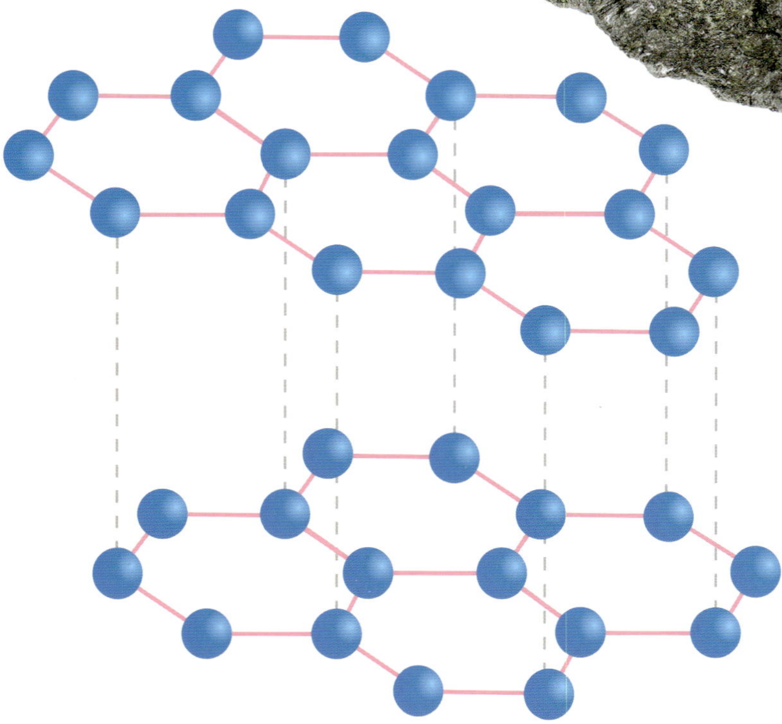

The carbon atoms in graphite are arranged in loose sheets.

Graphite Atoms

In graphite, the carbon atoms are structured in flat planes. Planes are like sheets of paper. Graphite is a very soft material.

Diamond Atoms

The carbon atoms in diamonds are structured in tetrahedrons. These shapes are similar to pyramids. This structure makes diamonds strong. In fact, diamonds are the hardest material in the world.

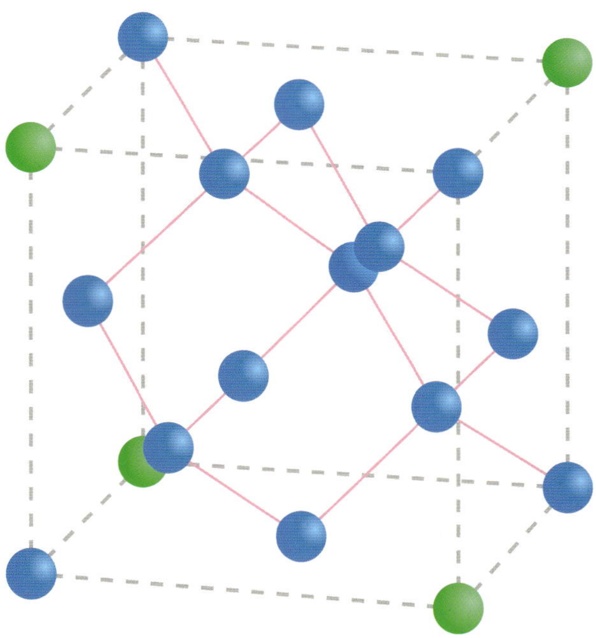

The carbon atoms in diamonds are tightly packed together.

How Are Minerals Used?

Everyone uses minerals. Minerals such as copper and mica are used in electronics. Selenite and gypsum are in toothpaste. People use salt to season food. Glass is made from melted quartz. Makeup and soap contain minerals such as talc.

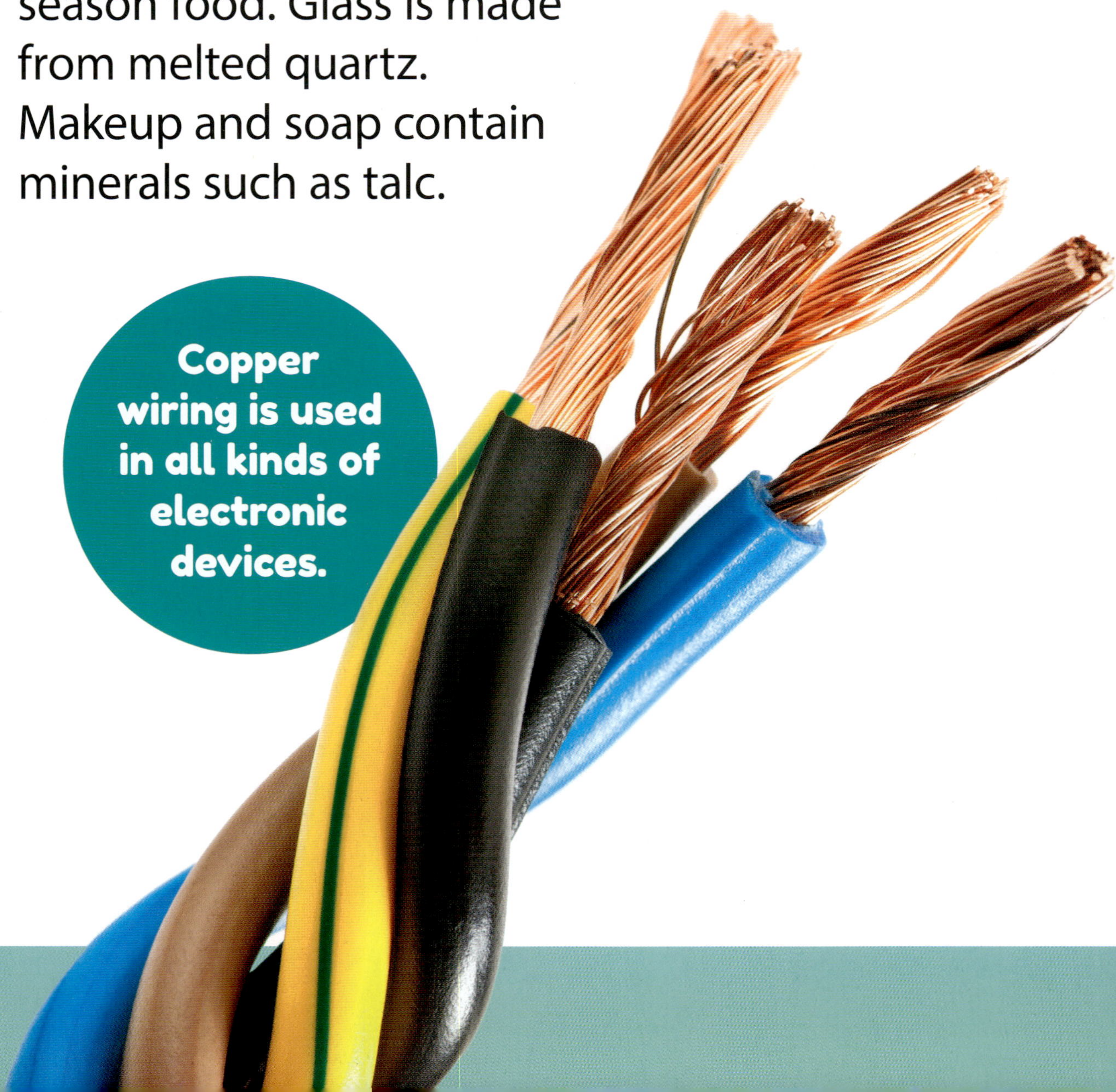

Copper wiring is used in all kinds of electronic devices.

FUN FACT!

People have mined emeralds for thousands of years.

Emeralds are popular minerals for jewelry.

Jewelry

Minerals are also used for jewelry. Diamonds, rubies, and emeralds are all minerals. People mine them all over the world.

A hardness test can help identify a mineral.

Mohs Hardness Scale

Geologists often find minerals while researching. Out in the field, it can be difficult to tell one mineral from another. Scientists use the Mohs Hardness Scale to classify minerals.

Testing Hardness

The Mohs Hardness Scale measures how easy it is to scratch a mineral. It is a scale from one to ten. One is the easiest to scratch. Ten is the hardest. Scientists can scratch a mineral against an object with a known hardness. That helps them identify the mineral.

Measuring Mineral Hardness

People can use the Mohs Hardness Scale to identify different minerals.

Alexandrite

Alexandrite is a mineral with an amazing color-changing trick. In sunlight, it looks blue or green. In indoor lighting, it looks more red or purple. Alexandrite is usually polished and cut as a gemstone.

Alexandrite's color-changing appearance can be seen in both its natural and polished forms.

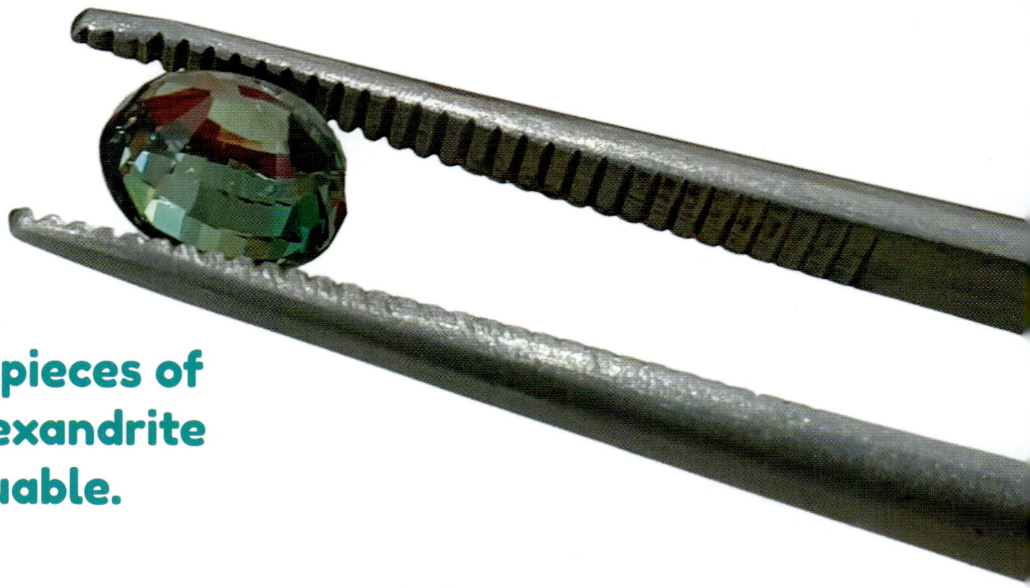

Even small pieces of polished alexandrite are valuable.

History

This mineral was named for Czar Alexander II. He ruled Russia in the 1800s. Alexandrite was discovered in Russia in 1830. It has also been found in Sri Lanka, East Africa, and Brazil.

The Difference Light Makes

Light sources have different colors. Some are more yellow. Some are more blue. This difference is what causes alexandrite to change color.

Many similar minerals belong to the amphibole category.

Amphibole

Amphiboles are a group of dark green or brown minerals. They can be opaque or translucent. Amphiboles are found all over the world. They are most often found inside igneous rocks.

Jade Decoration

Many cultures have used amphiboles to make decorations. When it is used for decoration, the material is called jade. The ancient Chinese made jade statues. The Aztecs once lived in Mexico. They used jade for jewelry.

People have been carving jade for thousands of years.

Beryl

Beryl comes in many colors. This includes pink, green, blue, and red. Many gemstones are types of beryl. Emeralds are dark green beryl. Pale blue beryl is called aquamarine. Morganite is a pink type of beryl.

Major producers of beryl include Brazil, South Africa, and the United States.

Morganite can be cut, polished, and used in rings.

Finding Beryl

Beryl is usually found in limestone and pegmatite rocks. Large crystals of beryl have been found all over the world. The largest was found in Madagascar. It weighed about 838,000 pounds (380,000 kg).

Calcite

Calcite comes in two forms. One is a big chunk that looks like a rock. These large pieces are usually gray or white. The other form of calcite looks like a crystal. The crystal form is smaller and usually colorless.

Animals and Calcite

Many animals use calcite to protect themselves. Oysters and other shellfish use materials from seawater to make calcite for their shells. Calcite is also found in the hard outer shells of some insects, such as leaf-cutter ants. Coral can also be made of calcite.

Calcite makes up part of oysters' shells.

Celestine

Celestine often looks like long, spiky crystals. It is normally white, pale blue, or yellow. It can also be colorless. The crystals can be opaque or translucent.

Celestine is found in sedimentary rocks.

The element strontium, found in celestine, helps create red fireworks.

Fireworks and Flares

This mineral is used in fireworks. It creates a bright red color when burned. Celestine is also used in flares. Flares are used for signaling in an emergency. People shoot them into the air. They explode with a bright flash.

Geodes

Geodes look like ordinary rocks. However, they hide a surprise. They are full of shiny minerals such as celestine, quartz, or amethyst.

Chalcopyrite looks similar to pyrite, but it is softer.

Chalcopyrite

Chalcopyrite looks mostly yellow. However, it has a tarnish. With the right lighting, chalcopyrite shows hints of green, blue, or purple. This mineral is always opaque.

Copper

The yellow color of chalcopyrite comes from copper. Pennies are coated with this element. Chalcopyrite is the most common source of copper. Copper was the first metal humans used to make weapons and tools. This makes chalcopyrite important in human history.

FUN FACT!

Copper wires are used in computers and televisions.

Copper is used to make cookware that heats quickly and evenly.

Chrysotile's fibers make it a dangerous mineral.

Chrysotile

Chrysotile is usually white, yellow, or green. It can be opaque or slightly translucent. It is soft. Chrysotile is a strange mineral. It sheds stringy fibers. They look like pieces of a torn-up cotton ball.

Asbestos

Chrysotile is a type of asbestos. Asbestos was once common in building products. It was used in floors, walls, and ceilings to control fires. But scientists learned that breathing in asbestos can cause illness. There are now many laws about using chrysotile.

Workers wear protective gear when removing asbestos from old buildings.

Cinnabar

Cinnabar is a bright red mineral. It sometimes looks like a dull rock. It can also be a shiny crystal. Water can cause it to break down. If it flows into another rock, it can color that rock red.

Cinnabar is often found near volcanic rocks.

Pure mercury is a metallic liquid that is very dangerous to human health.

Mercury

Cinnabar was once used as a dye. People in ancient China and Greece used it. However, this mineral is made of mercury. This element is harmful to eat or even touch. Today, it has been replaced with other substances for many uses.

Like other gemstones, corundum can be cut, polished, and used for jewelry.

Corundum

Corundum is a type of gemstone. It is opaque or translucent. It comes in many colors. Blue corundum is a sapphire. Red corundum is a ruby.

Impurities

Pure corundum is colorless. It is made of aluminum and oxygen. However, it can have impurities when different chemicals are present. The chemicals could be iron, chromium, or vanadium. Different impurities create the different colors.

The Stolen Star of India

The Star of India is a famous sapphire. It was found in Sri Lanka. The gemstone is as big as a golf ball. It was stolen in 1964. It was returned the next year.

Corundum can be found within other rocks.

Diamond

Diamonds are the most famous gemstones in the world. They are normally colorless. But they also come in other colors, such as pink or brown. They can be opaque or translucent. People often use diamonds for jewelry.

FUN FACT!

The Hope Diamond is a famous gemstone. Some people say it is cursed.

Major countries for diamond mining include Australia, the Democratic Republic of the Congo, and Russia.

Drill bits tipped with diamond can cut through tough materials.

The Hardest Mineral

No natural material is harder than a diamond. Diamonds can scratch all other minerals. This makes diamonds useful for drills. Diamond drill bits easily drill through other materials.

Dolomite contains the elements calcium and magnesium.

Dolomite

Dolomite is usually white or gray. However, it can be brown, red, or even pink. It can also be transparent.

Construction Material

This mineral is most often found in large pieces. It is usually found in sedimentary rocks. Dolomite is important for building. People often use it in concrete. This material is used to make roads. Concrete is also used to make buildings and even art sculptures.

Dolomite is an ingredient in some concrete.

Fluorite

Fluorite is most often purple. It can also be green, blue, white, gray, and even multicolored. Fluorite crystals are cube shaped. They are usually transparent. Ancient Romans used fluorite to make bowls and vases.

Fluorite often has vivid colors.

Fluoride in toothpaste helps protect teeth from decay.

Fluoride

Fluorite is the main source of fluoride. This chemical has many purposes. It is added to water and toothpaste to help strengthen teeth. It's also used to make cooking pans slippery so food doesn't stick to them.

Galena sometimes contains the element silver.

Galena

Galena is always gray and opaque. Pieces of galena break off in cube shapes. This mineral is denser than most other minerals.

Lead

Galena is the most important source of lead. Lead has been used to make many things. This element is found in computers, weapons, and even paint. However, lead can be dangerous if it enters a person's body. Because of this danger, many laws limit the mining of galena.

Mining for galena was common before people understood the dangers of lead.

Graphite breaks apart easily.

Graphite

Graphite is black and shiny. It is always opaque. Graphite is made of carbon that was once part of living things.

Soft or Strong

Graphite is one of the softest minerals. This is why it is used in pencils. However, graphite can also be strong. It can be formed into many shapes. People use it to make tennis rackets and running shoes.

Graphite Art

Artists use different kinds of graphite pencils. Varying amounts of clay are put into each pencil. Clay changes the pencil's darkness and affects how smoothly it draws.

Artists can choose from a variety of graphite pencils.

Gypsum

Gypsum is a white or clear mineral. It can be opaque or translucent. It is very soft. This mineral forms when salt water evaporates. The gypsum is left behind. It can be found near oceans. It can also be found in hot springs and caves.

Major countries for gypsum production include Spain, Thailand, and the United States.

Panels made of gypsum, called drywall, are used in the walls of houses.

Plaster

Gypsum is used in plaster. People use this material to cover walls and ceilings. Artists use it to form sculptures. Plaster is also used in casts for broken arms and legs.

Muscovite belongs to a group of minerals called micas.

Muscovite

Muscovite is usually white, silver, or colorless. It can be translucent or slightly opaque. Muscovite is often found in thin flakes or sheets. Flakes are tiny flat layers that easily fall off a crystal.

Making Windows

Sheets of muscovite can be very large. Long ago, people in India and Eastern Europe carefully carved sheets off muscovite crystals. They used the sheets for windowpanes before glass was invented.

Thin sheets of muscovite allow some light to pass through.

Olivine

Olivine is a category of green mineral. Olivine minerals have the same structures. The only differences are the elements they contain. They can be opaque or translucent.

Olivine usually contains either iron or magnesium.

Unlike many other gemstones, peridot appears in only one color.

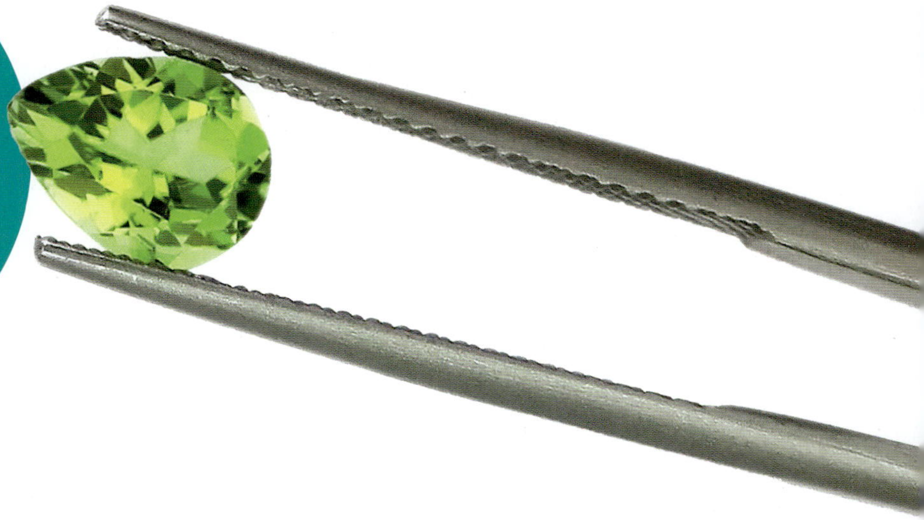

Peridot

Some olivines can be polished into peridot. This light green gem is sometimes called the evening emerald. Peridot was a treasured jewel for the ancient Egyptians. The largest peridot mines are still in Egypt.

Minerals in the feldspar group make up more than half of Earth's crust.

Potassium Feldspar

Potassium feldspar is a group of hard minerals that come in many colors. They can be white, red, orange, or brown. They are sometimes green. They can also be translucent or opaque.

A Durable Mineral

These minerals are usually found in igneous and metamorphic rocks. They are harder than glass. This means that potassium feldspar minerals do not weather as quickly as other minerals.

Potassium Feldspar

1 2 3 4 5 6 7 8 9 10

Mohs Hardness

Concrete and Pottery

Potassium feldspar minerals are used in concrete. They are also mixed with other minerals to create ceramic and porcelain. People use those materials to make pottery.

FUN FACT!

Green and blue potassium feldspar is used for jewelry.

Clay containing feldspar, along with iron and quartz, has been used to make stoneware jars.

Pyrite

Pyrite is a shiny gold-colored mineral. Its crystals often form squares. It is a heavy mineral. Hitting pyrite with other rocks makes hot sparks. Ancient people used pyrite to start fires.

Pyrite's name comes from the Greek word *pyr*, meaning "fire."

Gold is softer and denser than pyrite.

Fool's Gold

Pyrite is known as "fool's gold." People who find it sometimes think they have found gold. But real gold does not form in square shapes.

Pyroxenes are commonly found in igneous rocks, including basalt.

Pyroxene

Pyroxenes are a group of minerals that are dark green, brown, or black. Sometimes they are white or light green. These minerals are shiny like glass.

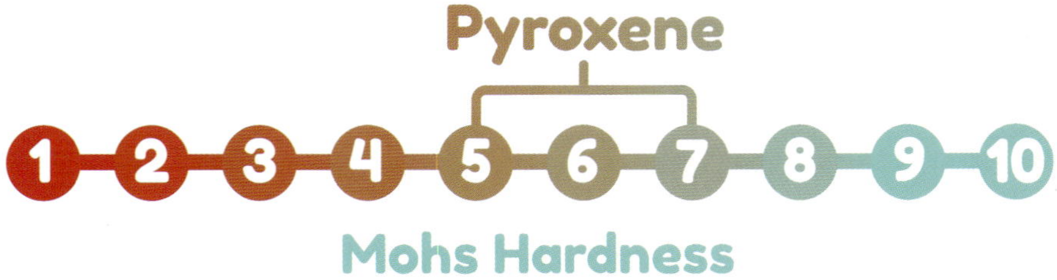

Jadeite Jewelry

Pyroxenes are often found after volcanoes erupt. As magma cools, a pyroxene begins to form. A kind of pyroxene called jadeite is shiny and comes in many colors. People use it for jewelry.

Both jadeite and the similar material made of amphibole are commonly known as jade.

Quartz

Quartz is one of the most common minerals. It is usually clear, but it also comes in many colors. Purple quartz is called amethyst. Other types of quartz are called rose, smoky, and milky. Quartz can be opaque or translucent.

Rutilated quartz often has a brownish or golden color.

Quartz

1 2 3 4 5 6 7 8 9 10

Mohs Hardness

A tiny quartz crystal helps some kinds of watches keep time.

Uses for Quartz

Quartz has many uses. This mineral is used in glass. The mineral's many colors make it popular for jewelry. It also has many uses in construction. Electronic devices use quartz too.

Chert

Quartz stones can also be opaque. They are called chert. Popular types of chert are jasper, onyx, and flint.

In the late 1700s, scientists realized spinel was a different mineral from ruby.

Spinel

Spinel comes in many colors. It can be blue, green, red, or purple. It can be polished into a gemstone.

Finding and Making Spinel

This mineral is found in many parts of the world. People mine it in South America, Asia, and Africa. Spinel has also been created in labs.

Spinel

1 2 3 4 5 6 7 8 9 10

Mohs Hardness

The Black Prince's Ruby

The Black Prince's Ruby is a famous gem. It looks like a ruby. But it is actually a large red spinel. It is on a crown worn by British royalty.

Minerals from Space

Spinel is often found in meteorites. Meteorites are rocks that fall from space. They contain many minerals.

The Black Prince's Ruby is on the front center of a British crown.

Sulfur

Sulfur is a yellow mineral often found around volcanoes. It can look like a rock or a crystal. It is opaque.

Changing Minerals

Sometimes sulfur starts as a different mineral. It may start as gypsum or anhydrite. These minerals sometimes contain bacteria. Bacteria are tiny organisms. They change the minerals into sulfur.

Sulfur is a very soft mineral.

Huge amounts of sulfur are used for manufacturing and farming.

Uses for Sulfur

Sulfur is often used for hardening rubber. It can also clean water. It is most commonly used to make fertilizers.

FUN FACT!

Sulfur smells like rotten eggs.

Talc has the softest rating of any mineral on the Mohs Hardness Scale.

Talc

Talc is a soft white mineral. It can be opaque or translucent. Talc has a strange texture. It feels soapy.

Softness and Strength

Though it is soft, talc is strong in other ways. It is not affected by heat. Acids or electricity do not damage it.

Everyday Uses

Talc is found in household objects. It's often used in paints. It can also be found in some makeup such as blush and eye shadow.

Talc is used in powdered beauty products, including eye shadow.

GLOSSARY

crust
The upper solid layer of a planet.

crystal
A repeating pattern of atoms.

evaporate
To turn from water into a gas.

mantle
The layer of Earth's surface below the crust, where magma is found.

opaque
Not allowing light to pass through.

organisms
Living animals or plants.

pegmatite
A type of igneous rock made of crystals.

porous
Having many small holes.

pressure
The physical force created when one object presses against another.

quarry
A place where people dig for rocks or minerals.

structure
The way a material is organized.

tarnish
A chemical change that affects the surface of a mineral.

texture
The way a material feels.

translucent
Allowing some light to pass through.

transparent
Allowing almost all light to pass through.

TO LEARN MORE

More Books to Read

Bassier, Emma. *Minerals*. Abdo, 2020.

Gieseke, Tyler. *The Rock Cycle*. Abdo, 2023.

Rocks and Minerals. DK, 2023.

Online Resources

Booklinks
NONFICTION NETWORK
FREE! ONLINE NONFICTION RESOURCES

To learn more about rocks and minerals, please visit **abdobooklinks.com** or scan this QR code. These links are routinely monitored and updated to provide the most current information available.

INDEX

PHOTO CREDITS

Cover Photos: Sebastian Janicki/Shutterstock Images, front (amethyst); Shutterstock Images, front (granite, malachite), back; Dima Moroz/Shutterstock Images, front (obsidian)

Interior Photos: Sebastian Janicki/Shutterstock Images, 1, 69, 122; Shutterstock Images, 3, 4, 6, 7, 8, 9 (top), 9 (bottom), 10, 12, 14, 19 (bottom), 20, 24 (top), 24 (bottom), 26, 28, 29, 31, 32, 33, 34, 37, 40, 42, 43, 44, 51 (bottom), 53, 56, 63, 65 (bottom), 66, 67, 70, 71 (bottom), 72 (bottom), 73 (left), 73 (right), 74, 75, 77 (middle left), 77 (right), 79 (top), 79 (bottom), 82, 83, 84, 85, 86, 87 (top), 87 (bottom), 88, 89, 91, 93, 95 (top), 96, 97, 105 (top), 105 (bottom), 107, 108, 109, 114, 118, 119 (top), 119 (bottom), 120, 121 (top); Evgeny Haritonov/Shutterstock Images, 5; Ronn Hook/Shutterstock Images, 11; TR Studio/Shutterstock Images, 13; Mendenhall Olga/Shutterstock Images, 15; Christine Minato/Shutterstock Images, 16; Ralf Lehmann/Shutterstock Images, 17; Aleksandr Pobedimskiy/Shutterstock Images, 18, 36, 106; Joseph Sohm/Shutterstock Images, 19 (top); Rattachon Angmanee/Shutterstock Images, 21, 22; Beth Ruggiero-York/Shutterstock Images, 23; Blue Ring Media/Shutterstock Images, 25; Sean Pavone/Shutterstock Images, 27; Armando Frazao/Shutterstock Images, 30; Boonchai Sakunchonruedee/Shutterstock Images, 35; Tyler Boyes/Shutterstock Images, 38, 41, 80, 110; Ekkaluck Sangkla/Shutterstock Images, 39; Richard Bradford/Shutterstock Images, 45; Anthony Heflin/Shutterstock Images, 46; Federico Rostagno/Shutterstock Images, 47; Jim Vallee/Shutterstock Images, 48; Ken Griffiths/Shutterstock Images, 49; Luciano Mortula/Shutterstock Images, 50; H. R. Photos/Shutterstock Images, 51 (top); Phil Cruz/Shutterstock Images, 52; Gary Whitton/Shutterstock Images, 54; Ale Rizzo/Shutterstock Images, 55; Galyna Andrushko/Shutterstock Images, 57; Mike Dotta/Shutterstock Images, 58; Robert Crow/Shutterstock Images, 59; Morphius Film/Shutterstock Images, 60; Polina Kuzovkova/Shutterstock Images, 61; Kim Briers/Shutterstock Images, 62; Planetary Visions Ltd/Science Source, 64; Andrea Danti/Shutterstock Images, 65 (top); Damian Pawlos/Shutterstock Images, 68; Albert Russ/Shutterstock Images, 71 (top), 92, 100; Bjoern Wylezich/Shutterstock Images, 72 (top), 95 (bottom), 102; Michael LaMonica/Shutterstock Images, 76; Minakryn Ruslan/Shutterstock Images, 77 (left), 77 (middle right), 78, 124; Mark Brandon/Shutterstock Images, 81; Krim Kate/Shutterstock Images, 90; J. Palys/Shutterstock Images, 94; Ground Picture/Shutterstock Images, 98; S. Phoophinyo/Shutterstock Images, 99; Soho A Studio/Shutterstock Images, 101; Denver Post/Getty Images, 103; Bruno Rodrigues B Silva/Shutterstock Images, 104; Yut Chanthaburi/Shutterstock Images, 111; Fokin Oleg/Shutterstock Images, 112; Sepia Times/Universal Images Group/Getty Images, 113; Marcel Clemens/Shutterstock Images, 115; Branko Jovanovic/Shutterstock Images, 116; The Natural History Museum, London/Science Source, 117; Print Collector/Hulton Archive/Getty Images, 121 (bottom); Alex Der Ant/Shutterstock Images, 123; Nik Merkulov/Shutterstock Images, 125